TROPE

SCIENCE PEOPLE

A Celebration of Our Diverse People of Science

Illustrated by

DAVID LEE CSICSKO

Text by

LINDY SINCLAIR

CONTENTS PAGE

CHEVY HUMPHREY

President and CEO,
Museum of Science & Industry

The Museum of Science and Industry, Chicago (MSI), is one of the largest and most popular science museums in the world. As its President and CEO, my goal is to inspire the inventive genius in everyone. In fact, for more than three decades my personal mission has been to make connections to science education available for people of all ages, across all communities. Wonderfully, David Lee Csicsko's *Science People* does just this. By highlighting the diversity of the scientific community around the world through his vibrant illustrations, David introduces readers to trailblazers across STEM fields, and shows us that scientists can come from anywhere, and look like anyone.

I am delighted that David has chosen to feature more than 50 scientists from such varied backgrounds and distinct fields, from around the world and throughout history. *Science People* includes a diverse mix of genders, races, and nationalities – from classic figures like Marie Curie and Albert Einstein to contemporaries working in the field now, such as Monkgomotsi Maseng and Jennifer Doudna, to Greta Thunberg, whose work is inspiring a new generation.

Representation is so important in realizing one's dreams. I believe if you can see it, you can be it. I was fortunate to have mentors to emulate, and I try to pay it forward by helping raise up the next generation in STEM fields. As a lifelong learner, I want to learn from everyone around me, no matter where I am. I know I've learned new things within the pages of *Science People*, and I trust you too will be entertained and informed by the profiles in this book. I hope that *Science People* inspires you to ask questions and think about your world in a new way. Because science really is for everyone – and is more important than ever.

CHEVY HUMPHREY

CHICAGO, ILLINOIS (USA)

ACKNOWLEDGEMENTS

This book is in celebration and memory of my cousin Dr. John Francis Csicsko, who gave me the keys that opened the doors to the big wide world of science. When I was a young lad, he gave me a small working microscope and later a chemistry set. These tools were about possibilities and wonder.

We live in a time where science is so important. My collection of portraits of people in science was a COVID art project. When the world shut down, I started drawing. The development of vaccines is saving lives as we journey together through these complex times. It's all about science and innovation, to help our health and to save our planet.

Science People is a visual history book featuring greats from the past as well as current scientists around the world. Read about them and marvel at the spark of an idea, tireless efforts, and dedication to a better world. It's been an education and a pleasure to work on this and see it become a book.

A big thank you to Anne Rashford, Monica Metzler, David Woody, Sabrina Tsao, Leslie Ramirez Bailey, Bob Weiss, Dr. Mary J. Pettersen, Jaci Zweig, Joy Pierce Mathews, Tracey Shafroth, Kyoji Nakano, Sam Landers, and David Syrek for supporting this project.

Also a thank you to my family who work in the medical community: Marylin Koonce, Mark Koonce, Bunny Csicsko Keebler, Janet Csicsko Mahoney, Michael Spafford, and Jenny Bailey. And to doctors and nurses everywhere.

DAVID LEE CSICSKO

SCIENCE
PEOPLE

HYPATIA

c. 355 – c. 415 AD

ALEXANDRIA (EGYPT)

HYPATIA

One of the first women in recorded history to study and teach math

Hypatia was one of the first women in history to be known as a mathematician and astronomer. She was born in Alexandria, Egypt, a city devoted to learning and home to some of the greatest thinkers of ancient times. The name *Hypatia* means "highest, supreme," and she certainly lived up to her name as the greatest mathematician and astronomer of her age.

When Hypatia lived, most women were limited to being wives and mothers, but Hypatia's father didn't want to keep his daughter at home. Theon was a well-known scholar himself, and he taught her math, astronomy, and philosophy (which includes what we now call science). Hypatia became an expert in all of them. She helped her father update textbooks with new information on geometry, algebra, and the sciences. Historians even suspect that some of Theon's writings were really by his brilliant daughter. She also wrote her own essays on math and philosophy, explaining earlier works and making them accessible to more people.

Hypatia had a gift for taking complicated ideas and breaking them into pieces that were easier to understand. She was known as a great teacher, and students came from all over Alexandria, and even from other cities, to learn from her. She was also an excellent public speaker. Crowds would come to listen to her talk about the ideas of Plato and Aristotle.

Although Alexandria had been a diverse city for centuries, tensions between religious groups were rising during Hypatia's lifetime. Some Christians thought her teachings went against their religious ideas. Hypatia was killed by a group of extremist Christians, and many of Alexandria's schools were burned after her death. Other intellectuals fled the city, and Alexandria's time as a center of learning came to an end.

Hypatia's death was a tragedy, but her life has become a symbol of education and independence.

AVICENNA (IBN SINA)

Philosopher-scientist who influenced medicine for centuries

Ibn Sina, known in the West as Avicenna, was a *polymath*, a person who knows a lot about many different subjects. He wrote about philosophy, art, math and languages. His students and followers called him *Al Shaikh Al Ra'ees*, or "master wise man." But he is most famous for his influence on early medicine, an influence that lasted for centuries.

Avicenna was a brilliant child. By the age of 10, he had memorized the Qur'an, the holy book of Islam. He started to study medicine at 13, and by age 21 was a well-known scholar and doctor. Sultans and princes around the Islamic world wanted him as their physician, and he spent a lot of time traveling.

He wrote about many subjects, but his most important work was his *Canon of Medicine*, five huge volumes that together were the most complete medical textbook of the time. It included all the medical knowledge Avicenna could collect, both from the Greeks and Romans and from Muslim sources. The *Canon* became the common medical reference across Europe for more than 400 years.

Even now, modern medicine might look very different without Avicenna's ideas. He introduced experiments to find the best treatments or to test the safety of medications. He discovered that some illnesses were passed from person to person, and he introduced the idea of quarantine to limit the spread of disease. He wrote about how diet and environment affect health. And he was one of the first to write about brain conditions like epilepsy, strokes, and hallucinations.

Avicenna made incredible advances in medicine, but he didn't stop there. His theories of motion later influenced Sir Isaac Newton. He was the first to classify simple machines like levers and pulleys. His theories about the speed of light and how mountains were formed were proven centuries later to be almost entirely correct. He was the greatest thinker of medieval times.

AVICENNA (IBN SINA)

980 – 1037

AFSHANA (UZBEKISTAN)

LEONARDO DA VINCI

1452 – 1519

ANCHIANO (ITALY)

LEONARDO DA VINCI

One of the greatest minds in history studied science to be a better artist

Leonardo da Vinci was one of the greatest artists in history. His paintings were more realistic and expressive than ever before. He was also a great inventor who created many designs hundreds of years before it was practical to build them, like the helicopter, army tank, solar power, even a mechanical robot. Many people think of art and science as opposites, but Leonardo used his intense curiosity about the world around him to make his art the best it could be.

Medieval science did not use experiments. Scholars based theories on ancient Greek and Latin writings, the Christian Bible, and their own common sense. Leonardo realized that he could learn more by carefully studying nature and asking questions. He also applied math to his observations in ways that had not been done before. Years before the scientific method was invented, Leonardo was a bridge between the old way of thinking and the careful steps scientists would follow in the future.

Leonardo dove into many fields of science in support of his art. He used experiments in anatomy to learn how the human body really worked and moved. He studied zoology and botany so that animals and plants in his paintings looked more realistic. He studied physics to understand how light reflects off a subject and where shadows fall. Chemistry helped him find exactly the right paint colors. He kept careful notes, filling thousands and thousands of pages with his sketches and writing.

What makes his discoveries even more remarkable was that Leonardo didn't have any formal schooling. He learned basic reading, writing and math as a child, but became an apprentice artist at 15. Instead, Leonardo followed his curiosity. He made connections between different subjects and was never afraid to try something new. He put no limits on his imagination, but knew art and science could work together to help us explore and understand our world.

NICOLAUS COPERNICUS

1473 - 1543

TORÚN (POLAND)

NICOLAUS COPERNICUS

Putting the sun at the center of our universe touched off the Scientific Revolution

For more than 2000 years, people thought Earth was the center of the universe. But sometimes planets move in unexpected ways, and ancient astronomers had a hard time explaining why they sped up or slowed down, or even seemed to move backwards. In the early 1500s, Nicolaus Copernicus looked for simpler explanations. His model moved Earth from the center of the solar system, to just another planet orbiting the sun.

Nicolaus was from a wealthy family, but his father died when he was young. His uncle, a Roman Catholic bishop, made sure Nicolaus got an education and a good job in the church. Nicolaus went to Italy in 1496 to study church law, but he was more interested in astronomy. He lived with the university's top astronomer, which is probably where Nicolaus first read books that doubted the Earth was the center of the universe.

After returning to Poland, he studied astronomy in his free time. In 1514, he wrote that the sun was the center of the universe. He sent his theory to friends and scholars but didn't publish it. The powerful Catholic Church taught that everything revolved around the earth, and he was afraid of how they might react. After re-checking his calculations for many years (the telescope had not yet been invented), Nicolaus finally published his theories. He was the first to name the correct order of Mercury, Venus, Earth, Mars, Jupiter and Saturn. He wasn't the first to think the sun might be the center of the universe, but his evidence was the best so far.

Not all of Nicolaus's work was correct, but he got enough right for later scientists to build on it. His book, published just two months before he died, started to change the way we understand our universe. Many historians believe this was the start of the Scientific Revolution, a time when many discoveries were made and the beginning of what we now call science.

GALILEO GALILEI

The "Father of Modern Science," whose natural curiosity and need to test his theories changed the way science is practiced

Galileo Galilei worked in so many areas of science and math that Albert Einstein called him the "father of modern science." As an astronomer and an engineer, he changed the way we understand our universe and how our own world works.

Galileo's father was a musician, but he wanted Galileo to study medicine to make a better living. Galileo was more curious about the world around him, though. He noticed a chandelier swinging back and forth and realized that the swings took the same amount of time whether they were large or small. His discovery led to the use of pendulums to keep track of time. He soon switched to study mathematics.

In 1609, Galileo heard about the invention of the telescope. He built one by reading only general descriptions and offered it to his government for spotting distant ships at sea. He kept making improvements until he could see the stars and planets. He discovered that Jupiter had four moons, Earth's moon had mountains, and Saturn had rings.

His discoveries proved that the sun was the center of the solar system, and that Earth revolved around it. The Roman Catholic Church taught that Earth was the center of the universe and ordered him not to talk about his ideas. Even though Galileo was religious, he refused to stay quiet. The church put him on trial in 1633 and punished him with house arrest.

Galileo was one of the first to show how the laws of nature could be understood with math, and proposed important theories about motion, force and gravity that we still use today. He combined experiments with math in a new way that is the basis of the modern scientific method. He did not look for evidence to confirm what he already thought, but formed (and changed) conclusions based on data gathered from careful study and testing.

GALILEO GALILEI

1564 – 1642

PISA (ITALY)

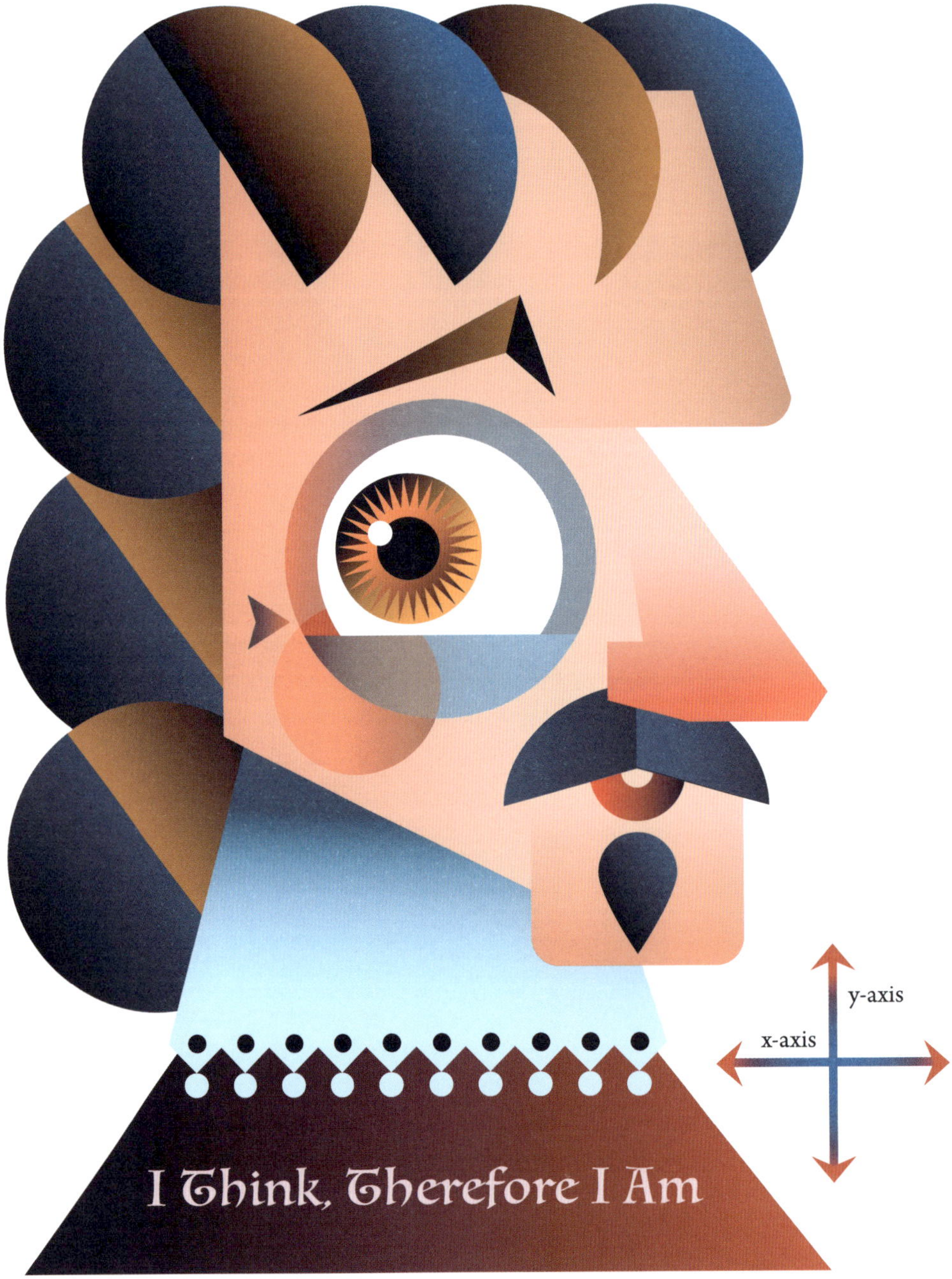

RENÉ DESCARTES

1596 – 1650

LA HAYE EN TOURAINE (FRANCE)

RENÉ DESCARTES

Father of analytical geometry and key influence on the Scientific Revolution

Before the 1600s, there was no such thing as a scientist; people who studied math and science were known as *natural philosophers*. René Descartes is often remembered as the father of modern philosophy – "I think, therefore I am" – but his influence on math and science might be even more important.

René's father wanted him to become a lawyer, but after law school René decided he'd rather travel, going to Paris and the Netherlands to study engineering and math. An extremely curious child, René decided he would rather study "the book of the world," as he called it, than bury his nose in law books.

In 1637, he published *Discourse on Method*, one of the most influential works in the history of science. Most books like this were written in Latin, read only by academic and religious elites, but René wrote *Discourse on Method* in French so that anyone "with good sense," as he put it, could read it and learn to think for themselves.

He thought it was time for a new kind of science, based on experiments and observation. His four rules helped form our modern scientific method: 1, Doubt everything. 2, Break all problems down into smaller parts. 3, Start by solving the simplest problems first. 4, Be thorough.

René applied his rules to problems in science and math with amazing results. He invented *analytical geometry*, showing how "unsolvable" problems in geometry could be solved by making them into simpler problems in algebra – two types of math that had not been linked before. Analytical geometry is widely used in engineering, space exploration, and aviation – many modern fields would not exist without it.

René's ideas were used by later thinkers, including Isaac Newton, to develop calculus and modern physics. He had transformed these fields forever. The Scientific Revolution was a time of enormous change and advancement, and René's writing was at the heart of it all.

ISAAC NEWTON

Scientific genius who invented calculus and changed physics forever

Isaac Newton was born too early and wasn't expected to survive. Not only did he live, he grew up to be one of the greatest figures in the history of science – Albert Einstein kept an image of Isaac on the wall of his study! It is hard to imagine today's math and science without his discoveries.

As a boy, Isaac's mother pulled him out of school to work the family farm. It was a disaster – Isaac read or built mechanical projects instead. Despite his genius, he wasn't a top student in college; just like on the farm, he spent more time reading advanced math and science than he did on his assignments.

The Great Plague of 1665 closed his university, and Isaac spent two years mostly in isolation. With nothing to do but study, this is when Isaac came up with many of his most important ideas. He saw an apple fall from a tree and wondered why it fell straight down (it didn't actually fall on his head). He wanted to know why the moon orbits the earth instead of floating off into space. And he needed math formulas that didn't exist to figure it all out.

Isaac invented the first reflecting telescope, a design still used today. He discovered that white light is made of all the colors of the rainbow. But it took many years of work before Isaac made his other theories public. His book *Principia* might be the most influential scientific work ever published, and it made him famous all over the world. His ideas about force and gravity helped explain nearly every motion in the universe. The math formulas he invented to prove his theories became known as *calculus*, a type of math that makes advanced engineering and science fields possible.

Many of today's scientists still name Isaac Newton as an inspiration. He once said that "If I have seen further than others, it is by standing upon the shoulders of giants." But many scientists and historians say that his shoulders might be the greatest of all.

ISAAC NEWTON

1643 – 1727

WOOLSTHORPE, ENGLAND (UK)

MARIA SIBYLLA MERIAN

1647 – 1717

FRANKFURT, HESSE (GERMANY)

MARIA SIBYLLA MERIAN

Pioneered the study of insects and turned them into fine art

Everyone knows that caterpillars turn into butterflies, but it took a patient observer to prove that two creatures who look so different were in fact the same. Maria Sibylla Merian did this as one of the first *entomologists* – scientists who study insects – and then turned her discoveries into fine art.

Maria's stepfather was a painter, and as a child she collected plants and insects as models for her own painting. Most artists back then painted an insect by itself, but Mary spent hours watching it to add its life cycle or what it ate. The word *ecology* wasn't invented until the late 1800s to describe the science of how living things exist in their environments – but Maria studied insect ecology almost 200 years earlier.

Instead of using preserved insects, Maria made her kitchen a laboratory. She raised mostly butterflies and caterpillars, and sometimes she would sit up all night just to see a newly formed butterfly escape its cocoon. In 1679, she published *The Wondrous Transformation of Caterpillars*. There was no other book like it, showing how insects transformed over their lives and how they fit into the wider natural world.

In 1699, Maria took her youngest daughter on one of the first-ever scientific expeditions, a dangerous 5000-mile voyage to South America. She got to know the area's enslaved Africans and Indigenous people, who brought her insect specimens and taught her the uses of local plants. The book Maria published when she returned home was full of new insect species and an important reference for both artists and scientists.

Maria's careful observations led to many discoveries about plants and insects. For example, many people believed that insects came from mud, which she proved was untrue. Scientists across Europe used her work in their own studies of nature. By linking art and science, she changed both forever.

ALEXANDER von HUMBOLDT

Explorer who showed us that all of nature is interconnected

Alexander von Humboldt was the greatest explorer-scientist of his time. His travels convinced him that nature was one connected web, and that humans could upset ecosystems and change the climate. This idea was revolutionary in the 1800s, but we know that he was right, and Alexander is considered by some as the first environmentalist.

Born to a wealthy family, Alexander was expected to study politics and finance. Instead, he fell in love with botany. There wasn't a lot of plant variety around Berlin, so Alexander dreamed of traveling to faraway places to study natural life. He eventually got permission from Spain to visit Central and South America.

For five years, Alexander traveled more than 6000 miles on land and water. He spent months in tropical forests and climbed the highest mountain in Ecuador. Alexander found the magnetic equator, a huge discovery that let him measure latitude and longitude more accurately and make better maps. He studied how an area's geography and its plants and animals were connected. He measured temperature and weather patterns. When Alexander returned to Europe, he spent many years publishing his data and his theories.

Alexander believed that science should be shared, and he wrote most of his books so that non-scientists could understand them. He organized one of the first conferences where scientists from different countries could trade ideas and information. He helped get support for stations in Canada, South Africa, and Australia to study magnetic storms – one of the first international scientific collaborations.

Alexander's curiosity was never satisfied. "People often say that I'm curious about too many things at once," he said, "But can you really forbid a man from harbouring a desire to know and embrace everything that surrounds him?" His idea that all of nature is connected changed the way people in the West saw the world. His discoveries are still used by scientists in many different fields today.

ALEXANDER VON HUMBOLDT

1769 – 1859

BERLIN (GERMANY)

MARY ANNING

1799 – 1847

LYME REGIS, ENGLAND (UK)

MARY ANNING

Her fossil discoveries helped lay the foundation for modern paleontology

Mary Anning was born in a seaside town in England. Her family was very poor, and her father made extra money by combing the beaches for fossils to sell to tourists. By the time Mary was 5 or 6, she was helping her father hunt for fossils among the sand and rocky cliffs. There were plenty to find – Lyme Regis is part of what scientists call the Jurassic Coast, where fossils are still being discovered.

Mary's father died when she was 11, and she continued to sell fossils to help her family. She didn't have much schooling, but she could read, and taught herself about geology and anatomy from books. She dissected fish and other small sea animals to better understand her finds.

About a year after her father's death, Mary's brother Joseph found an odd skull. For months, Mary carefully dug around the rest of the skeleton. People did not yet know about extinction and thought it must be a monster crocodile. It is now known that Mary and Joseph discovered the first skeleton of an ichthyosaur, an extinct marine reptile.

In 1823, she found the first complete plesiosaurus – another extinct sea reptile. The skeleton was so strange that there were rumors it was fake. Even some scientists did not believe it was real. Today, scientists still use Mary's skeleton when studying plesiosaurs.

Male scientists bought her fossils, but almost never gave her credit. The Geological Society of London would not even invite her to meetings where they studied and debated the fossils she uncovered. They did, however, start to take Mary's finds more seriously, and many visited Lyme Regis to discuss them with her.

Mary continued to uncover clues to the earth's past. Her work sparked popular interest in geology and paleontology and provided key evidence for the theory of extinction. She is now recognized as an enormous influence on the history of science and how we understand our planet's development.

CHARLES DARWIN

1809 – 1882

SHREWSBURY, ENGLAND (UK)

CHARLES DARWIN

Adventurer and biologist who transformed what we know about the natural world

Charles Darwin's father, a doctor, wanted his son to follow in his footsteps, but the sight of blood made Charles sick to his stomach. Instead, he fell in love with natural history. He collected plants and insects as a boy and stayed curious about the seemingly unlimited variety of life on Earth. His theories gave us a new understanding of how the natural world changes over time.

After completing his studies at Cambridge, Charles jumped at the chance to join an expedition to South America aboard the HMS *Beagle*. For five years, Charles collected plant, animal and rock samples, amazed by the many species he saw. He started to notice that birds and other animals seemed to adapt to their environments. Finches, for example, had different beak shapes on different islands, depending on the seeds or insects available there.

Back in England, Charles shared his samples with other scientists and began to study his notes about his travels (more than 1700 pages!). By 1938, Charles had developed a theory called *natural selection* – that animals who can adapt to their environment live longer and have more young. They pass on their adaptations to their children, which over many years causes evolution of the species, or even new species altogether.

Charles knew his ideas went against religious teachings of the time, and even some beliefs of his own. He wanted more evidence before he made his ideas public. It took him 20 years to finally publish *On the Origin of Species*. The book was controversial, but quickly became a worldwide bestseller. Each time the book was reprinted, Charles added more evidence and responses to his critics. In his later years, Charles also wrote about how human biology changes over time.

Charles's theories themselves have evolved as scientists build on his work. But his close observations and careful research upended the way we understand the natural world and our place in it.

ADA LOVELACE

1815 – 1852

LONDON, ENGLAND (UK)

ADA LOVELACE

Gifted mathematician celebrated as the first computer programmer

Young Ada Lovelace dreamed of flying, studying birds and testing materials for wings. Combining science and imagination, Ada had the perfect mind to dream of the machines of the future - and became the first computer programmer nearly 100 years before the digital computer was invented.

Ada had an unusual education for women of her time, with tutors in math, logic, and science. Ada's mother, a mathematician herself, thought that studying art and poetry might make Ada run wild like her absent father, the famous poet Lord Byron.

When Ada was 17, she met inventor Charles Babbage, who showed her a model of his Difference Engine, a hand-cranked mechanical calculator made up of numbered wheels. Ada was amazed and requested the blueprints so that she could see how it worked. She and Charles continued to write to each other about math and science for more than 15 years.

Charles started to think about a new project called the Analytical Engine, a machine for much more advanced calculations. In 1842, a mathematician published a paper in French about the Analytical Engine idea. Ada decided to translate the paper into English, but she added her own notes and thoughts. When she finished, her notes were three times as long as the original paper, and they were incredible! Ada described how the device could work and showed how punched cards could generate a specific sequence of numbers - now considered the first computer program. But she also imagined uses for it far beyond math calculations.

It took almost 100 years for computing to catch up to Ada's theories. Her notes were republished in 1953 as computer science was just getting started, and people finally started to understand her brilliance. An early programming language was named for her in the 1970s, and the second Tuesday in October is now Ada Lovelace Day, an international day to celebrate women's contributions to math, science, and technology.

LOUIS PASTEUR

Modern medicine and food safety would be impossible without his work on germs

It seems strange now, but for many years people thought that diseases and spoiled food were caused by "bad air." Louis Pasteur was one of the first to prove that it was germs, tiny organisms invisible to the eye, that were the culprit. It is hard to imagine what food safety and medicine would be like without his discoveries – and how many millions of lives have been saved because of his work.

Louis was a talented artist, but after he realized how little money most artists made, he became a teacher of physics and chemistry instead. When he tried to help a business owner figure out why his product kept going bad, he discovered it was full of bacteria. He thought the bacteria might be causing the spoilage – back then, people thought that food spoiling *created* bacteria.

Louis knew that boiling wine killed bacteria, but it also ruined the taste. By slowly trying warmer and warmer temperatures, he found a temperature that killed bacteria without harming the liquid. If you've ever read a milk carton, you've probably heard of his process, called *pasteurization*. Cheese, juice, honey, and many other products are pasteurized today to keep them fresh and keep us from getting sick.

Once he understood that germs caused food to spoil, Louis started to study how they might also cause diseases. He saved the French silk industry by finding the organisms attacking silkworm eggs and creating a way to separate healthy eggs from diseased ones that is still used today. He discovered that exposure to a weakened version of a disease could protect against it and created the first vaccines for anthrax and rabies.

Louis was trained as a chemist and knew nothing about the human immune system. Yet by carefully observing, testing theories, and never giving up, Louis made one of the most important discoveries in medical history, a revolution in biology that changed the world.

LOUIS PASTEUR

1822 – 1895

DOLE (FRANCE)

THOMAS EDISON

"The Wizard of Menlo Park" made modern life possible with his inventions and business sense

Thomas Edison had imagination and curiosity, but he wasn't a good student. Hearing problems meant that he missed a lot of what teachers said, so he was bored and easily distracted. He was a great reader, though, and read everything he could. He would become one of the most famous inventors of all time and change modern life forever.

Thomas quit school at age 12 to work on the railroad, then a few years later started work as a telegrapher. At the time, telegraphs were dots and dashes on paper. As they started to use sound, Thomas invented devices to make up for his hearing difficulty and improve the equipment. By the age of 22, he became a full-time inventor and entrepreneur.

In 1876, he moved to a lab in Menlo Park, New Jersey, that he built with his father. The very next year he created the phonograph, or record player, which could record and play back sound. People were so amazed they thought it was a trick. Newspapers started calling him the "Wizard of Menlo Park."

For 50 years, scientists had been trying to replace gas-powered lights. Others had invented an electric-powered light bulb, but their designs weren't practical. Thomas was the first to create one that could be manufactured and sold, then he developed a power station to deliver electricity to businesses and homes. Though Nikola Tesla soon invented a better way to create and deliver power, Thomas had introduced the modern age of electricity.

Thomas held more than 1000 patents and created the first industrial research lab. He invented one of the earliest movie cameras and an electric car battery. But the key to Thomas's success was that he wasn't afraid to try new things - and fail. When an invention didn't work, Thomas moved on. He thought of failures as chances to learn: "I have not failed 10,000 times," he said, "I've successfully found 10,000 ways that will not work."

THOMAS EDISON

1847 - 1931

MILAN, OHIO (USA)

NIKOLA TESLA

Genius engineer whose dreams powered our future

Nikola Tesla claimed he was born during a lightning storm. Whether or not it's true, it's a good story for the man whose power system made it possible to bring electricity to homes and buildings across the country. He's not as well-known as his rival Thomas Edison, but his inventions may have changed our lives even more.

Nikola always wanted to be an engineer, but he left college without graduating and worked in Budapest and Paris before emigrating to the United States in 1884. He worked as an engineer for Thomas Edison in New York but quit after six months – Thomas offered him $50,000 if he improved one of Thomas's designs, but when Nikola succeeded, Thomas told him he had only been joking.

Nikola knew his electrical system, called *alternating current* (AC), was better than Thomas's direct-current (DC) system. He found a business partner and by 1893, when his AC system lit up the World's Fair in Chicago, he had won the "War of the Currents." A few years later, General Electric built the first modern power station at Niagara Falls with Nikola's help and designs.

The electric motor Nikola invented is one of the most useful inventions of modern times, which is why Tesla Motors is named for him. He was also a pioneer in radio communication and dreamed of a wireless society – an idea far ahead of his time. He proposed using alternative energy like wind and geothermal decades before most scientists realized the problems with fossil fuels.

Nikola spoke eight languages and could do advanced calculus in his head. Some of his ideas were so fantastic that people called him a magician; when he demonstrated a remote-controlled boat, people thought it was fake. Some ideas were too far ahead of their time to be built at all. Others may have been better at business, but Nikola's brilliant and imaginative mind dreamed of the future.

NIKOLA TESLA

1856 – 1943

SMILJAN (CROATIA)

GEORGE WASHINGTON CARVER

Groundbreaking agriculturist whose ideas about sustainability were years ahead of his time

Most people connect George Washington Carver with peanuts. He invented or promoted hundreds of ways to use peanuts – including soap, plastics and face cream – but his work was much more important than one plant. George's ideas about sustainability were ahead of his time and would not sound strange to environmentalists today.

George's early life is a remarkable story. He was born enslaved and orphaned as a baby. Freed after the Civil War, he left the plantation when he was around 11 years old, working odd jobs and picking up whatever education he could as he slowly made his way north. In his 20s, he became the first Black student at the Iowa State Agricultural College.

After graduation, George went to teach at the Tuskegee Institute in Alabama. He was excited to direct Tuskegee's new agriculture department and help African American farmers. Cotton was the South's biggest crop, but George knew that it used up soil's nutrients quickly. At first, he tried to get farmers to change crops, but he learned that most poor Black farmers had no choice – nothing made as much money as cotton.

Instead, George started to teach them to farm more cheaply. He encouraged them to grow their own food. He taught them to make more of their needs from the land itself – compost as fertilizer, paint from clay and soybeans. He found that peanuts could be alternated with cotton to keep soil healthy and provide nutrition for farm families.

George knew the health of the land and its people were connected many years before this kind of thinking was common. His inventiveness made him famous around the world; foreign governments wanted his help with their own agricultural problems. When he died in 1943, President Franklin Roosevelt said, "The world of science has lost one of its most eminent figures."

GEORGE WASHINGTON CARVER

1864 – 1943

DIAMOND, MISSOURI (USA)

MARIE CURIE

First woman to win a Nobel Prize, and the only person ever to win in two different science fields

Maria Skłodowska took after her father, a math and physics teacher. Though she was a top student, the University of Warsaw did not admit women. Maria read and worked with tutors until she made it to the Sorbonne in Paris in 1891, where she studied physics, chemistry and math. She tried to find teaching work in Poland, but Krakow University also rejected her because of her gender, so she returned to Paris.

Marie, as she was now known, was inspired by the discovery of x-rays and of a strange energy coming from the element uranium. She studied whether other elements might also have this energy. Soon she was joined by her new husband Pierre Curie, also a physicist, who put aside his own work to help her investigate substances that were "radio-active" (a word she invented). She discovered that radioactivity came from the atoms themselves, a breakthrough that rewrote ideas about atoms.

In 1903, Marie became the first woman to win a Nobel Prize. Originally, only Pierre had been nominated, but he insisted that her contribution was at least as important as his. When Pierre died in an accident, Marie took over his teaching position and became the first woman to teach at the Sorbonne. In 1911, Marie won a second Nobel Prize for discovering two new elements, polonium and radium. To this day, only four people have won two Nobel Prizes – and no one else has won in two different scientific fields.

Marie traveled the world to raise money for more research on how radioactive substances could be used, especially in medicine. During World War I, she invented an x-ray unit that could be taken directly to the battlefield to help wounded soldiers. She was the first major female scientist to get full credit for her work, and her bravery and brilliance has inspired generations of girls and women to follow in her footsteps.

MARIE CURIE

1867 – 1934

WARSAW (POLAND)

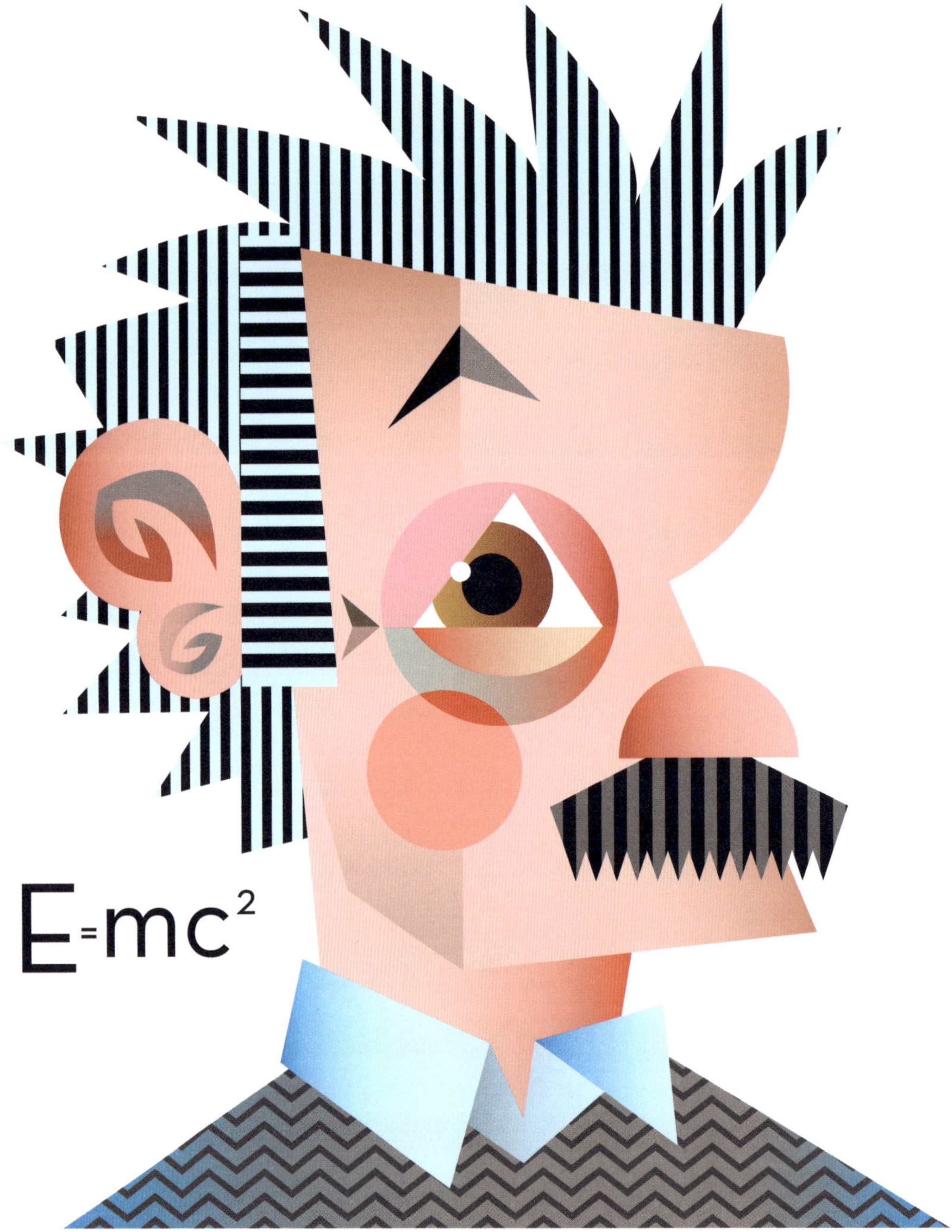

ALBERT EINSTEIN

1879 – 1955

ULM, BADEN-WÜRTTEMBERG (GERMANY)

ALBERT EINSTEIN

Reinvented physics for the modern age

Albert Einstein saw a compass when he was 5, and he was fascinated by the invisible force that made the needle move. Once he learned that math and geometry could help him figure out the wonders of the world, there was no looking back. If the word "scientist" makes you imagine a man with a bushy mustache and wild white hair, you're not alone. Albert was one of the most influential scientists of the 20th century, and his face has become a symbol of science – and how it can change everything.

Albert wanted to teach math and physics, but he had trouble finding a teaching job. He worked instead as a patent clerk, which gave him a lot of time to think. In 1905, Albert wrote four papers that turned science on its head. He offered the first mathematical proof that atoms existed, and even how to measure them. He showed that light was a stream of particles instead of a single wave – this idea, called the *photoelectric effect*, won him a Nobel Prize in 1921. His theories of relativity showed that ideas about gravity, light, and even space and time needed to be reconsidered.

At just 26, Albert reinvented physics and started new areas of study. He finally became a professor, teaching in Zurich and Prague before returning to Germany. (He moved to the United States in 1933 to escape the Nazis.) He became an international superstar in 1919 when he correctly predicted that stars would appear to be in a slightly different place during a solar eclipse, proving one of his theories of relativity.

Scientists still work on questions that Albert made possible, from how the universe started to what happens in a black hole. For hundreds of years, astronomers looked for mathematical reasons for what they observed in the world. Albert flipped that around – he followed his calculations wherever they led, even to the biggest questions in the cosmos.

ALEXANDER FLEMING

Discovered penicillin and changed the future of medicine

Before antibiotics, even a scratch could land you in the hospital if the wrong bacteria got into your body. Ancient Egyptians applied moldy bread to infected wounds, so the idea that some substances kill harmful bacteria was not new, but until Alexander Fleming discovered penicillin, there were no treatments for many common infections.

Born on a farm, Alexander came to London at age 13 to live with a stepbrother who was a doctor. He planned to become a surgeon, but a job in the new field of bacteriology put him on a different path. He started working with Sir Almroth Wright, a pioneer in the field of vaccines.

We think of modern scientists in sterilized labs, but Alexander kept his workspace a little messy on purpose. He always looked for the unexpected, hoping nature would surprise him. Neither of Alexander's two great discoveries were planned – and if he hadn't been so observant, he could have missed both.

Alexander was getting over a cold in 1921 when a drop of mucus splashed into a sample of bacteria he was growing. Rather than throw it away, he waited to see what would happen, and days later found that the bacteria had dissolved. He had discovered an enzyme in body fluids that attacks bacteria, and it convinced Alexander that living organisms could create material to fight infection.

A few years later, a sample was accidentally left out when Alexander went on vacation. When he studied it, he not only found that a mold called *Penicillium* had gotten into it, but that the bacteria around the mold was destroyed. Penicillin eventually was used during World War II to save the lives of many injured soldiers. Alexander shared a Nobel Prize in 1945 with two scientists who figured out how to purify penicillin so it could be made into a reliable medicine. It finally gave doctors a tool to treat infections, revolutionizing modern medicine.

ALEXANDER FLEMING

1881 – 1955

DARVEL, SCOTLAND (UK)

MICHIYO TSUJIMURA

1888 – 1969

OKEGAWA, SAITAMA PREFECTURE (JAPAN)

MICHIYO TSUJIMURA

Groundbreaking biochemist known for her research on green tea's nutritional benefits

Michiyo Tsujimura's father, a school principal, encouraged education for Michiyo and her six siblings. One of Michiyo's college teachers was Kono Yasui, who had recently become the first Japanese woman to earn a doctoral degree in science. Michiyo had always been interested in plants, and her professor inspired her to do scientific research of her own.

Hokkaido Imperial University, one of Japan's top schools, did not admit women. To continue her research, Michiyo had to work - unpaid - as a laboratory assistant. She started there in 1920, studying the nutrition of silkworms. She moved to another university in 1922, but a devastating earthquake destroyed the lab. When Michiyo evacuated mid-experiment, she took one of the lab's most valuable instruments with her, showing how dedicated she was to her work!

Michiyo went to work at a large research institute with Dr. Umetaro Suzuki, a pioneer in nutritional science. With another scientist, Michiyo discovered that green tea was a good source of vitamin C. After she published her results, there was a dramatic increase in Japan's green tea exports to North America. When a western scientist claimed there was no vitamin C in green tea after all, Michiyo studied his experiment and found his mistake - he boiled the tea for so long he destroyed the vitamin C. When the tea was made in the traditional Japanese way, the vitamin C remained.

Michiyo continued to find more chemical elements of green tea, including *catechin* and *tannin*, two of the components that give green tea its taste. In 1932, she published "On the Chemical Components of Green Tea" and became the first woman in Japan to earn a doctorate in agriculture.

Michiyo became a professor herself, teaching at several Japanese universities. A few years later, she patented a way to extract vitamin C crystals from plants. In 1956, she was awarded the Japan Prize of Agricultural Sciences to honor her research.

ISOBEL WYLIE HUTCHISON

"The Quiet Explorer" brought new understanding of the Arctic's plant life and Indigenous peoples

Isobel Wylie Hutchison was a botanist, but also a painter, a writer, a filmmaker, and above all, one of the greatest Arctic explorers of her time. She loved plants, but botany was really Isobel's ticket to independence and an escape from society's expectations of women.

Isobel was born at Carlowrie, her family's castle, and had a fairy-tale childhood until her father died when she was 10. Isobel dealt with her grief by taking long walks. After two brothers died when she was in her 20s, her hikes got longer – up to 100 miles. Though she was already a self-taught expert, Isobel studied botany at Studley College in her late 20s. She decided that marriage was not for her and wanted to make her own way in the world.

After a Mediterranean trip with an overprotective companion, Isobel decided she would rather travel alone. In 1925, Isobel walked over 250 miles across Iceland – a trip everyone said was impossible, with no maps and no guides at the time. She brought home some rare plants and enough material for a 30-page feature in *National Geographic*. She set her sights on Greenland next, and the Royal Horticultural Society helped her get permission to study the native plant life.

Isobel spent more than two years learning from Greenland's native people, filming documentaries, and collecting plants and seeds for botanical gardens and private collectors. In the 1930s, Isobel trekked through Alaska and the Aleutian Islands to gather dozens of plants never before seen in the UK. Museums and gardens across the country wanted samples of the plants and handmade goods she brought back.

She gave lectures, wrote books, and published many articles. Her love of travel was in everything she did, from her paintings to the languages she spoke (at least seven). Her daring adventures through some of the world's harshest terrain brought valuable understanding of the Arctic's people and plant life.

ISOBEL WYLIE HUTCHISON

1889 – 1982

WEST LOTHIAN, SCOTLAND (UK)

JOAN BEAUCHAMP PROCTER

1897 – 1931

LONDON, ENGLAND (UK)

JOAN BEAUCHAMP PROCTER

Devoted herpetologist who changed the way zoos study and house reptiles

Joan Beauchamp Procter kept snakes and lizards as pets as a kid. She brought a Dalmatian lizard to the table with her at mealtimes and even once brought a crocodile to school! Joan was an excellent student, but she was sick a lot. Her illness kept her from going to college, but she continued to study the animals she loved.

She started writing to George Boulenger, a famous zoologist at the British Museum. He was so impressed with her passion and knowledge that he invited her to work as his assistant. She presented her first scientific paper to the Zoological Society of London when she was only 19. When George retired a few years later, Joan took over as reptile keeper – the first woman to have that job.

In 1923, Joan helped design a new aquarium at the London Zoo. That same year, at only 26 years old, Joan was hired as their curator of reptiles. A few years later, she designed a new reptile house, which opened in 1927 and is still in use (a scene from the first Harry Potter movie was filmed there). It was the most advanced reptile enclosure in the world at the time, using Joan's ideas about environments that would reduce stress on the reptiles in her care – and would later be copied by zoos all over the world.

Joan published in many scientific books and journals, becoming known as an expert in herpetology, or the study of reptiles and amphibians. Her close relationship with the reptiles let her identify new diseases and invent new veterinary techniques, often with tools she herself designed. She was the first person to describe the giant Komodo dragon in captivity; she even tamed one, walking it around the zoo to greet visitors. She discovered the peninsula dragon and had a snake and a tortoise named after her.

Joan had health problems her whole life and died at only 34 years old. But by then, she was one of the most famous and respected herpetologists in the world.

JANAKI AMMAL

Pioneering botanist and advocate for India's natural diversity

Janaki Ammal inherited curiosity about the natural world from her father, who loved to garden and wrote two books about birds. At the time, fewer than 1% of women in India could read, but with her family's encouragement Janaki earned two degrees before winning a scholarship to study in the US. She became the first Indian woman to earn a PhD in botany at an American university and returned to India as an expert in plant cytology, the study of a plant's cells and genetics.

India had grown sugar for more than 2500 years, but by the 1930s most came from other countries who grew sweeter types of sugarcane. Janaki studied how different plants could be combined to form new hybrids, and eventually created a new, sweeter sugarcane plant that grew well in India's climate.

Janaki faced discrimination as an unmarried woman from a caste, or social class, that was looked down on. She left for London, where she co-wrote the *Chromosome Atlas of Cultivated Plants*, recording and organizing genetic information for 100,000 plants. This book is still used by plant scientists today.

Janaki returned to India in 1951 to collect and study native plants. She traveled to some of India's most remote areas, looking for medicinal plants or learning how agriculture was developing in different regions. She soon became an environmental activist, working to protect India's biodiversity and native plants.

As an expert in India's plant life, she worried about environmental damage caused by cutting down too many forests to grow food or produce power. She spoke out against a plan to build a hydropower dam in the rainforests of the Silent Valley, and helped get the project canceled. Today, Silent Valley National Park is one of the last undisturbed forests in the country.

In recognition of her contributions, Janaki was the first woman scientist to be awarded the Padma Shri, one of the highest non-military awards in India.

JANAKI AMMAL

1897 – 1984

THALASSERY, KERALA (INDIA)

STAR MAPPING VATICAN NUNS

Born late 1800s

LOMBARDY REGION (ITALY)

STAR MAPPING VATICAN NUNS

"Lady computers" who mapped the position of nearly half a million stars

Women have always contributed to science, but many have been overlooked. Four Italian nuns worked behind the scenes on one of the most ambitious science projects in history, only to have their names forgotten. But Sisters Regina Colombo, Concetta Finardi, Luigia Panceri, and Emilia Ponzoni are now known as the women who spent 11 years on the night shift to help us learn more about the heavens.

In 1887, scientists met in Paris to plan the Celestial Map Project, hoping to map every star using 22,000 glass plates containing pictures of the night sky. The work was divided among almost 20 institutions across the US and Europe, including the Vatican, headquarters of the Roman Catholic Church.

After its project director died in 1894, the Vatican didn't make much progress. In 1906, a new director with experience in astronomy finally took over, but he had never done the kind of measurements and calculations needed. He learned many observatories relied on women to do that hard work, almost always for low pay. One observatory even called them "lady computers" for all their calculations to figure out star positions.

The Vatican recruited two nuns but quickly realized they needed more. From 1910 to 1921, these four women stayed up all night recording the position and brightness of a staggering number of stars – a total of 482,215! Only about half of the original observatories completed their assignment, but the Vatican was one of them thanks to Regina, Concetta, Luigia and Emilia.

Photos of the nuns have appeared in books about the history of astronomy, but their identities were lost until 2016, when a priest at the Vatican Observatory unearthed their names while organizing archives. The Celestial Map Project ended in 1966 after recording nearly 5 million stars. Even with today's advanced technology, astronomers have discovered that comparing their findings to star positions recorded 100 years ago still reveals valuable information.

ENRICO FERMI

1901 – 1954

ROME (ITALY)

ENRICO FERMI

Brilliant physicist who fathered the Atomic Age

Enrico Fermi was 14 when his older brother died suddenly. Enrico dealt with his grief by studying extra hard. He discovered an interest in quantum physics – a new field at the time that studies things as small as an atom (or even smaller). No one could predict the nuclear-sized impact that Enrico would have on the world.

After high school, Enrico applied to a competitive university for gifted students. Not only did he win a scholarship, he did so well that the university put him straight into graduate school. He skipped regular undergraduate studies and finished his doctoral degree in four years!

Enrico found that almost every element (the most basic substances, made of only one type of atom) could be made radioactive by splitting the atoms apart. Radioactive atoms give off particles and energy. In 1938, Enrico won the Nobel Prize. Italy had started to pass anti-Jewish laws, and because his wife was Jewish, Enrico was afraid for his family's safety. After accepting his prize in Sweden, he and his family fled to the US instead of returning to Italy.

Once in the US, Enrico discovered how to create a nuclear chain reaction, with the energy from split atoms causing nearby atoms to split, too. A chain reaction could release a huge amount of energy. He designed the first nuclear reactor and showed how these chain reactions could be started and controlled.

His discoveries helped lead the world into the Atomic Age. Nuclear energy can be destructive (nuclear bombs) or useful (generating electricity), but there is no doubt that it changed the world.

Enrico's enormous contributions to physics have been honored in many ways. Element #100 in the periodic table is named *fermium* after him. Super-tiny particles that scientists call the "building blocks of matter" are *fermions*. And one of the highest science and technology awards given by the US government is named the Enrico Fermi Award in tribute to his brilliance.

HELIA BRAVO HOLLIS

The "Queen of Cacti" was a world-famous expert on Mexico's native plant life

Growing up in Mexico City, Helia Bravo Hollis took long Sunday walks with her parents and loved the cacti and other plants in the parks they visited. Few botanists studied cacti at the time, so she couldn't find much information about them. Helia herself changed that by becoming the world's greatest expert on cactus and succulent plants.

Her family wanted her to study medicine, but Helia preferred the natural world. Biologist wasn't even considered a career yet, so she forged her own path and became the first certified biologist in Mexico. She was a hard worker, first publishing in the *Revista Mexicana de Biología* (Mexican Biology Journal) when she was still a teenager. At first she studied protozoa, tiny single-cell organisms, but soon she came back to the plants she loved as a kid.

By 1930, Helia was traveling the country collecting samples for the new Institute of Biology. The plant family *cactaceae* is one of the most diverse in Mexico – more than 700 species. In 1937, Helia published *Las Cactáceas de México (The Cacti of Mexico)*, which proved she had become one of the leading experts in her field.

In 1951, she helped start the Mexican Society of Cactology. She and other members thought that creating a botanical garden was important for representing and sharing Mexico's native plants. They convinced authorities to grant them land, and in 1959, Helia became the garden's first director. She was so passionate about the garden that she once paid workers herself during a strike.

Over her long career, Helia wrote three books and more than 160 articles. Her contributions were so important that there is an entire category of cacti (*Heliabravoa*) and at least six species named after her. Perhaps most fitting, the Helia Bravo Hollis Botanical Garden in Puebla, Mexico, is home to many endangered cactus species and teaches visitors about their importance to the ecosystem.

HELIA BRAVO HOLLIS

1901 – 2001

MEXICO CITY (MEXICO)

JACQUES-YVES COUSTEAU

1910 – 1997

SAINT-ANDRÉ-DE-CUBZAC (FRANCE)

JACQUES-YVES COUSTEAU

Legendary ocean explorer and filmmaker

Jacques Cousteau dreamed of flying. He joined the French Navy to train as a pilot, but a terrible car accident ended that dream. He had always loved to swim, and started swimming in the ocean to rebuild his strength. The first time he swam underwater with goggles, he knew he had found another way to fly.

Jacques wanted to stay underwater longer than diving equipment at the time allowed. He worked with an engineer to invent the Aqua-Lung, the first scuba gear to automatically supply air when a diver inhaled. Later he came up with other inventions, including a tiny submarine called a diving saucer, waterproof cameras, and the first habitat that let researchers live underwater.

During World War II, Jacques experimented with underwater filming and research. After the war ended, he joined other divers and scientists on the first shipwreck expedition to use the new automatic diving gear, which completely changed underwater archaeology. A few years later, in 1950, he converted an old military ship named the *Calypso* into a research ship and started exploring waters around the world.

For the next 40-plus years, Jacques educated millions of people about the beautiful and mysterious world under the sea with his books, films, and TV shows. He swam with whales, interacted with octopuses, and hitched a ride with giant turtles. He explored the mountains of the deeps. He brought attention to ocean research and showed us why the ocean was worth protecting.

Jacques didn't start out as an environmentalist, but the more he learned about the oceans, the more committed he became to saving them. He helped convince world leaders to ban most commercial whaling and campaigned against dumping nuclear waste into the Mediterranean Sea. He believed that we have a responsibility to future generations to guard the natural world so that they will be able to enjoy it. He is now remembered as one of the ocean's greatest defenders.

ISABELLA AIONA ABBOTT

The "First Lady of Limu" was one of the world's greatest experts on Pacific marine plants

As a child, Isabella Aiona Abbott learned about *limu* (Hawaiian for edible algae and seaweed) from her mother. They foraged along the beaches to collect limu to use in traditional Hawaiian dishes. Isabella fell in love with Hawaii's diverse plant life, a passion she followed to become one of the world's greatest marine botanists.

Isabella studied botany through school and became the first native Hawaiian to earn a science PhD in 1950. When her husband got a job at Stanford University, she moved to California. It was almost impossible for a woman to become a professor then, but Isabella studied the seaweeds of her new home and eventually started lecturing at Stanford. She was such an amazing teacher, and published so many papers and field guides, that Stanford skipped the usual steps and made her a full Professor of Biology in 1971 – the first women and first person of color with that title.

Over her amazing career, Isabella wrote eight books and more than 150 research papers. Her book *Marine Algae of California*, over 800 pages, is still considered the best guide to Pacific coast algae. She discovered more than 200 species – some named after her, such as the red algae Abottella.

Isabella realized her Hawaiian ancestors' knowledge about marine plants was quickly disappearing. She studied algae's historical uses and interviewed elders to collect what they knew for future generations. She moved back to Hawaii in 1982 and helped start the University of Hawaii's major in ethnobotany, the study of how people of a particular place use their native plants.

Her passion for seaweed wasn't limited to work. She was an excellent cook, and often brought a cake made with seaweed to potlucks. *Gourmet* magazine even wrote about her. She loved to invent recipes and bring food to class for her students to try. She was named a Living Treasure of Hawaii in 1997, after a lifetime spent preserving Hawaii's botanic heritage.

ISABELLA AIONA ABBOTT

1919 – 2010

HANA, HAWAII (USA)

ROSALIND FRANKLIN

1920 – 1958

LONDON, ENGLAND (UK)

ROSALIND FRANKLIN

Captured the double helix structure of DNA

Rosalind Franklin's family valued education and public service – values Rosalind would more than live up to as one of the 20th century's most brilliant scientists. She was an exceptional student, even doing math problems for fun. By age 15, she knew she wanted to be a scientist.

She studied physics and chemistry, earning her PhD researching properties of coal. She then spent several years in Paris learning x-ray diffraction, a complex way to create 3D images of crystals for the study of molecules. She not only perfected her technique, but found new ways to use it.

She returned to London in 1951 to join a team trying to figure out the structure of DNA. Scientists already knew it formed the basic blocks of the human body, but understanding what DNA looked like could help them learn so much more. One of Rosalind's x-ray diffraction pictures, now known as Photograph 51, proved DNA was shaped like a double helix – imagine a twisted ladder.

Without telling her, one of Rosalind's coworkers showed Photograph 51 to scientists at another university also working on DNA. James Watson and Francis Crick created a model of DNA using that photo as key evidence – but did not give her credit when they won a Nobel Prize for their work.

Rosalind spent the rest of her life studying viruses, along with more work on coal. She unlocked the structure of RNA in the tobacco mosaic virus – ironically, correcting earlier work by James Watson. Her work became a model for other biologists trying to break the genetic code. She did interesting work on the polio virus and traveled the world as an expert on virus structure.

Rosalind wanted her research to make a difference in the world, and it still does. Because of her work, researchers can use tools like DNA sequencing to study new viruses. The structure of DNA is considered one of the most important scientific discoveries of all time, and Rosalind played a key part.

TETSUYA "TED" FUJITA

1920 – 1998

KITAKYUSHU (JAPAN)

TETSUYA "TED" FUJITA

"Mr. Tornado" shaped twentieth century meteorology

Tetsuya Fujita was fascinated by storms as a kid in Japan. His PhD was a study of typhoons, giant tropical storms in the Pacific and Indian Oceans. He wanted to understand how storms functioned, and how they formed. His work changed meteorology, the science of weather.

He was a 24-year-old physics professor when the atomic bomb hit Nagasaki, killing more than 50,000 people. Tetsuya was asked by the government to survey the damage. With careful measurements and observations, he noticed patterns in the destruction that helped him figure out where the bomb was when it exploded. He used his new knowledge to study the destruction caused by large storms. In the early 1950s, he was invited to the University of Chicago to work on improving weather forecasting. He moved to the United States and adopted the name Ted.

Ted gathered evidence like a scientific Sherlock Holmes. He walked for miles around towns and fields after a tornado, taking photographs and measurements. He flew over damage in small planes to get a wider view and used satellite pictures in new ways to observe weather. Over time, "Mr. Tornado," as he was often called, created a scale to classify tornado strength – a scale still used today, more than 50 years later.

Just as important was his discovery of microbursts, sudden and powerful air currents that last only a few minutes minutes but can cause incredible damage. After scientists could not explain a 1975 airplane crash, Ted went into detective mode, looking at radar and interviewing pilots that had safely landed around the same time. He suspected that if a sudden wind gust was small enough, it might not show up on the airport's weather monitors.

His theories were dismissed by other scientists, but he never backed down. He spent years collecting evidence, and eventually proved his theory. Airlines changed safety regulations to protect from microbursts, saving thousands of lives.

MARIE THARP

Geologist whose maps forever changed our understanding of the ocean floor

Little Marie Tharp liked to tag along as her father collected soil samples and made charts for the US Department of Agriculture. She didn't think she could follow in his footsteps; science was considered "men's work." But during World War II, so many men joined the military that women were recruited into jobs they'd been denied before. Marie studied geology and math, then moved to New York as one of the first women at the Lamont Geological Observatory.

At the time, not much was known about what the bottom of the ocean looked like; scientists mostly assumed it was flat. Marie teamed up with Bruce Hazeen to map the ocean floor, carefully calculating and drawing underwater details by hand using data from sonar readings. Rather than a muddy plain, the data showed a seafloor covered in mountains and canyons, just like on land.

Mapping the Atlantic Ocean, Marie discovered a huge valley between long mountain ridges, so deep that she kept re-checking her calculations looking for a mistake. She thought it might be evidence that the underwater mountains were pulling apart - a theory called *continental drift*, the idea that continents move over long periods of time. Mainstream science thought that idea was absurd, and it took her almost a year to convince even Bruce to take her seriously.

Marie's maps eventually showed 40,000 miles of underwater ridges circling the earth. Scientists who tried to prove them wrong only found more evidence of the continental drift theory, causing a revolution in geology.

Marie continued with the first serious attempt to create a map of the entire ocean floor. Published in 1977, her World Ocean Floor map is still widely used today. In 1997, the Library of Congress recognized Marie as one of the greatest cartographers of the 20th century. She gave us a first look at a part of the planet we'd never seen before, and new understanding of the ground under our feet.

MARIE THARP

1920 – 2006

YPSILANTI, MICHIGAN (USA)

EUGENIE CLARK

The "Shark Lady" was one of the first scientists to use scuba diving to study fish behavior up close

Eugenie Clark spent most of her life in and around the water. She learned to swim before she was 2 and took her last dive at age 92. In between, she changed the way we understand ocean life.

Marine biologist was an unusual career choice for a girl from New York City in the 1940s. Eugenie fell in love with the ocean early. Her mother was born in Japan, an island nation where the sea holds great cultural importance. Eugenie would wander the New York Aquarium for hours as a child, and never stopped wanting to learn all she could about the ocean and its creatures. Even when she faced prejudice for her gender or her Japanese heritage, her love of the ocean kept her studying and exploring.

Eugenie was a pioneer in the use of scuba diving for ocean research. At the time, she was one of the only ichthyologists – scientists who study fish – to use diving to study living specimens up close. Over her long career, she discovered three new species of fish, wrote several books, and traveled the world sharing her knowledge with others. She appeared in 50 TV specials or documentaries. One even won her an Emmy Award! In all her work, she spoke out strongly for taking care of our oceans and shorelines.

But her nickname was the "Shark Lady" for her work with the most feared – and to Eugenie, most misunderstood – creatures of the deep. She proved that sharks could be trained, and that not all sharks need to swim constantly to breathe. Her discoveries totally changed the way scientists understand shark behavior and biology.

Eugenie's legacy will ripple through ocean research for generations. Her research added so much to our knowledge of the ocean, but just as important, she inspired schoolchildren around the world to follow in her wake.

EUGENIE CLARK

1922 – 2015

NEW YORK, NEW YORK (USA)

LEON LEDERMAN

1922 – 2018

NEW YORK, NEW YORK (USA)

LEON LEDERMAN

Explorer of the universe's tiniest particles and advocate for better science education

Particle physics is one of the hardest branches of science. It studies matter and energy at the tiniest, most impossible-to-see levels, and many basic rules of physics don't even apply. Leon helped develop particle physics with his inventive experiments, and his work gave us a deeper understanding of the subatomic world.

Leon started studying chemistry, but found physics more exciting, with new discoveries all the time. Not long after earning his PhD, he discovered a new kind of *neutrino*, the most common type of particle in the universe but one of the hardest to detect. (Neutrinos don't interact with other particles – trillions are passing through your body right now!) Scientists now think that neutrinos are key to learning how the universe changes over time. Leon was one of the first to show that studying these tiny particles and learning about the cosmos were related.

Serious stuff, but Leon was famous for his sense of humor. He knew his work confused most people; when he won the Nobel Prize, he said in his acceptance speech that "the Two Neutrinos sounds like an Italian dance team." He wanted to make science less "scary" and show how much fun it could be.

Leon also thought a lot about how science is taught. In many high schools, students learn biology, then chemistry, with physics last. That order was backwards to Leon. He thought the sciences should be taught in an order that links them together: physics first, to learn about atoms. Chemistry next, since atoms team up to form molecules. Biology would come last, as molecules combine into living things.

The high school he helped start near Chicago is now one of the top US schools for math and science. After he retired, he led an organization that trained elementary school teachers on better ways to teach math and science. Leon worked in areas of science that are hard for us to imagine, but he always looked for ways to share the joy of discovery.

TU YOUYOU

Her anti-malaria medicine has saved millions of lives, but she went unknown for years

Artemisinin is a drug to treat malaria, a disease that has killed people for thousands of years. It is one of the most important drug discoveries of the 20th century, but for decades, no one knew who had discovered it, only that it had come from China.

In 2015, the world learned the real story when Tu Youyou was awarded the Nobel Prize in Physiology or Medicine. She was the first Chinese winner in that category, and the first female Chinese citizen to win a Nobel Prize, period.

Tu's parents thought education was extremely important, but her studies were interrupted for two years when she got really sick with tuberculosis. When she returned to school, she knew she wanted to study medicine and find cures for diseases like the one she'd recovered from.

Her path to greatness was unusual – Tu doesn't have a medical degree or a PhD. She went to pharmacology school, then spent three years studying traditional Chinese medicine. In 1969, she became the leader of a secret government project to find a cure for malaria. It was a huge task; scientists around the world had already tested over 240,000 compounds, and none were successful.

They looked for clues in ancient Chinese medical texts, and found a reference to a plant called *qinghao*, or sweet wormwood, used around 400 AD to treat the kind of fever that is usually a symptom of malaria. After a lot of trial and error, they created a recipe that worked in mice and monkeys, using chemicals taken from wormwood leaves. Tu tested the new drug on herself first before starting tests on human patients. All of them got better.

It was several years before her work was shared outside of China, and at first it was published anonymously. Tu has always given credit to her research team, as well as the ancient healers who inspired her. But her passion and refusal to give up resulted in a drug that now saves lives all over the world.

TU YOUYOU

Born 1930

NINGBO, ZHEJIANG (CHINA)

JANE GOODALL

Born 1934

LONDON, ENGLAND (UK)

JANE GOODALL

Made us rethink what it means to be human by studying our closest animal relatives

Inspired by her favorite books, Jane Goodall dreamed of someday living among wild animals in Africa. Classmates laughed at her, but her mother said that if she worked hard and never gave up, she just might make it happen – and she did. What she found there changed the relationship between humans and animals forever.

In 1957, Jane went to visit a friend in Kenya, where she went to work for an anthropologist searching for fossils and clues to human evolution. Though Jane didn't have a college degree, she impressed her new boss so much that he sent her to Tanzania to study chimpanzees. Instead of spying on the chimpanzees from far away, Jane spent months gaining their trust so that she could study them up close.

Jane soon saw the chimpanzees making and using their own tools. Scientists were forced to rethink ideas about how people were different from animals; making tools had been the dividing line. Jane also discovered that chimpanzees were omnivores, not herbivores, and that chimpanzees, our closest living relatives, had social interactions and even personalities that were much more human-like than expected.

In the 1980s, Jane also became an activist. At first she was concerned that the number of chimpanzees (and the size of their natural habitat) was getting smaller. She quickly saw that solving environmental problems also meant working with local communities on poverty, health care, and education. Since then, Jane has traveled the world to raise money and awareness for conservation efforts.

Jane changed how we think about how people fit into nature. Instead of being separate from it, we are part of a whole system. What does it mean to be human? How are we linked to the creatures we share our planet with? Jane realized that animals, the environment, and people are all connected, and that to make the world better for one, we need to improve it for all.

CARL SAGAN

1934 – 1996

NEW YORK, NEW YORK (USA)

CARL SAGAN

Groundbreaking astronomer and America's "most effective salesman of science"

Five-year-old Carl Sagan wanted to know about stars. What were they made of? Why did they shine? He asked his parents, friends, librarians – but none of their answers were enough. For the rest of his life, Carl tried to understand the cosmos, and to help others understand it, too.

Carl loved science. He was president of his high school chemistry club and set up a home laboratory. He even spent his free time reading science fiction. When he found out that astronomer was an actual job, there was no looking back. He graduated high school at 16 and went to the University of Chicago, one of the only schools that would let in a student so young.

Degrees in hand, he taught at Harvard and then Cornell, two top universities. He quickly got attention with research showing how harsh the environments of Venus and Mars would be for any kind of life. American space exploration was just getting started, and Carl helped design and manage NASA's voyages to Venus, Mars and Jupiter. He helped ready astronauts for trips to the moon.

But Carl didn't only want to do science – he wanted to share it. He was so good at explaining ideas to non-scientists that *Time* magazine called him "America's most effective salesman of science." He wrote hundreds of articles and more than two dozen books – including one that won a Pulitzer Prize. He was a frequent talk show guest. He co-created and hosted *Cosmos*, a public television series shown around the world.

For years, Carl was the most famous scientist in America. Other scientists sometimes didn't take him seriously because of his popularity, but he made major contributions to astronomy. He helped build the fields of planetary science and exobiology, the study of whether there might be life on other planets. Carl never stopped searching for what might live among the stars, and he wanted to bring everyone along for the ride.

WANGARI MUTA MAATHAI

1940 – 2011

NYERI (KENYA)

WANGARI MUTA MAATHAI

Scholar and social activist who linked environmental issues and human rights

Wangari Muta Maathai was born in a rural area in central Kenya. She forged her own path from an early age, winning a scholarship to attend college in the United States. People in the US were becoming more aware of pollution and environmental issues in the mid-1960s, and Wangari got her first taste of environmental activism. She earned two degrees and returned to Kenya, which had recently become independent after many years of British colonial rule.

At the University of Nairobi, Wangari became the first female scholar in east and central Africa to earn a PhD. She became the university's first female professor, not only teaching veterinary anatomy but becoming the head of the department.

In the late 1970s, women across Kenya were reporting streams drying up. They had to walk farther to gather wood for fuel and building, and food was getting harder to grow. Wangari had the idea that women all over the country could plant trees to help the soil and store rainwater. In 1977, she founded the Green Belt Movement to help rural communities rebuild their ecosystems and act on climate change.

Wangari connected conservation and environmental science to the fight for good government, women's rights, and poverty reduction. The Green Belt Movement spread to other African countries, and has contributed to the planting of more than 51 million trees. The Pan African Green Belt Network tries to educate world leaders about the need for environmental improvements, and Wangari addressed the United Nations many times.

Speaking out often put her in conflict with businesses and the government, but her bravery won her support all over Kenya. She was elected to the National Assembly in 2002 with an incredible 98 percent of the vote. In 2004, she became the first African woman to win the Nobel Peace Prize. Wangari's love of science led her to dream of a better world - and to fight for it.

STEPHEN HAWKING

World-famous theoretical physicist who explored the cosmos from his wheelchair

Stephen Hawking was born exactly 300 years after the death of Galileo and died on Albert Einstein's birthday. As one of the greatest minds in physics, Stephen would surely appreciate this connection with other historical giants in his field. Like Galileo and Einstein, Stephen turned physics upside-down with theories that sound impossible or bizarre but become accepted as evidence piles up proving them right.

Stephen explored black holes, places in space where gravity pulls so hard that not even light can escape. He combined Einstein's theories with complicated math in totally new ways. He convinced scientists that instead of just a theory, black holes might actually exist in space. (He was right.) He disproved old theories – even some of his own – and added new ones. He tried to answer questions about where the universe came from, how it was formed, and how it worked.

While Stephen's brain roamed the universe, his body grew weaker. At age 21, he found out he had a rare disease called ALS that would paralyze his muscles. Doctors told him that he had only a few years to live. Instead, Stephen lived to age 76 and became a world-famous expert in his field. When he got too weak to move or even speak, he used computers that could read small facial movements to communicate (being British, he liked to joke about his voice synthesizer's American accent).

Stephen was known for his sense of humor and daredevil attitude. He became a best-selling author, a guest star on *The Simpsons* and *The Big Bang Theory*, and the subject of a movie about his life, *The Theory of Everything*. He traveled the world teaching and giving lectures. ALS took away Stephen's ability to move his body but did nothing to stop his brilliant mind. "I want to show that people need not be limited by physical handicaps," he said, "as long as they are not disabled in spirit."

STEPHEN HAWKING

1942 – 2018

OXFORD, ENGLAND (UK)

KATIA KRAFFT

1942 – 1991

GUEBWILLER (FRANCE)

MAURICE KRAFFT

1946 – 1991

MULHOUSE (FRANCE)

KATIA & MAURICE KRAFFT

Daredevil couple who advanced the study of volcanoes by getting up close and personal

As a teenager, Katia Conrad decided she wanted to be a volcanologist, someone who studies volcanoes and why they erupt. There were hardly any volcanologists at the time, and they were almost all men – but Katia didn't care. For her 18th birthday, she convinced her parents to take her to Mt. Etna in Italy, one of the world's most active volcanoes.

At the University of Strasbourg, she met Maurice Krafft, who was also obsessed with volcanoes. They teamed up in work and in life; when they got married, they went to an island with three active volcanoes for their honeymoon. Their house became a two-person volcano research center.

Katia and Maurice were often the first on the scene of an active volcano. Volcanoes are unpredictable, so many scientists were afraid to study eruptions in person. Not the Kraffts. They got as close as they could, even camping within feet of molten lava! No one else did this kind of up-close investigation, so they had to invent some of their own tools. They created special helmets to protect their heads from flying rocks, and Katia invented a tool called a chromatograph to measure volcanic gases.

They were a good team; Katia took pictures and collected samples while Maurice recorded video and took notes. Over 20 years studying volcanoes, it's thought that they visited half of all the volcanoes on earth! The information from their expeditions made huge advances in the study of volcanoes.

Over time, they got frustrated by how little was done to protect people living near volcanoes. They made films to educate people about the risks and dangers, and Katia worked with local governments on safety and evacuation procedures.

Katia and Maurice died on an expedition to Japan's Mount Unzen in 1991, when the volcano suddenly erupted. They knew the dangers better than anyone, and their work cost them their lives. But it has saved countless others all over the world.

JOCELYN BELL BURNELL

Born 1943

LURGAN, NORTHERN IRELAND (UK)

JOCELYN BELL BURNELL

Astrophysicist who discovered a new type of star and advocate for women in science

Jocelyn Bell's parents believed that girls had a right to an education when that opinion wasn't common. Her father was an architect who had helped build the local planetarium, and he encouraged her interest in science. When Jocelyn complained that her school sent girls to cooking classes while boys went to science lab, her parents "hit the roof." By the end of the semester, Jocelyn was ranked at the top of the science class.

Jocelyn wanted to study the stars. She read her father's astronomy books over and over until she understood them. But being a woman in science was tough. At college, male classmates heckled her in the lecture hall. She was one of only two women in her graduate program. In fact, she was afraid that Cambridge had admitted her by mistake, so she worked extra hard.

For two years, she helped build a massive telescope that used radio waves to pick up on very distant objects. She was assigned to look for *quasars*, or young galaxies. She had to analyze 100 feet of chart paper a day – boring work. One day she found a signal she couldn't explain. She kept taking measurements and found a consistent pulse coming from deep space. Soon she found another.

Pulsars – short for "pulsating radio stars" – became one of the century's most exciting astronomy discoveries. They are like outer space lighthouses, more regular than an atomic clock. The study of pulsars has led to many other discoveries. In 1974, Jocelyn's advisor and a colleague were awarded the Nobel Prize. Jocelyn was not included, but she was thrilled to know her discovery was so important.

Her work was not forgotten. In 2018, Jocelyn was awarded a Breakthrough Prize, nicknamed "the Oscars of science." She donated the entire $3 million prize to fund physics scholarships for female, non-white, and refugee students. Her hope is that these young scientists will become the role models she never had.

SUE HENDRICKSON

Self-taught marine archaeologist and fossil hunter

As a kid in Indiana, Sue Hendrickson often walked with her head down, looking at the ground. "People said, 'Look up. Smile!'" she says. "Now, I realize I was born to look for things and just didn't know it."

She moved to Florida as a teenager and learned to dive, collecting tropical fish to sell as pets. She tagged along with other divers to explore shipwrecks. On a visit to the Dominican Republic, she saw an insect perfectly preserved in amber and was fascinated to learn that it was millions of years old.

Sue read everything she could about fossils and fell in love with paleontology. She started collecting amber fossils for museums and universities. On a dig for ancient whale bones in Peru, she met paleontologist Peter Larson. He invited her to dig for dinosaur bones with his organization in the Black Hills of South Dakota.

In August 1990, their last planned day in the Black Hills, the team's truck got a flat tire. While others went for help, Sue stayed with the truck and investigated some dark shapes on the side of a cliff. They were fossilized bones – a lot of them! It took almost three weeks of careful digging to free them all. The bones turned out to be the largest, most complete *Tyrannosaurus rex* skeleton ever discovered.

The team nicknamed their find "Sue" in honor of her sharp eyes and instinct to investigate. It took two and a half years to put Sue together, and it was eventually purchased by the Field Museum in Chicago. Scientists have learned about the way *T. rex* looked and moved by studying this incredible discovery.

Sue kept exploring and diving, following her adventurous spirit to Egypt and Cuba. She was part of the expedition that discovered the Spanish ship *San Diego* off the coast of the Philippines nearly 400 years after it sank. She now lives a quiet but busy life on an island just off the coast of Honduras – and still loves to dive.

SUE HENDRICKSON

Born 1949

CHICAGO, ILLINOIS (USA)

BILL NYE

Inventor and educator dedicated to making science fun

Bill Nye loved bikes and airplanes as a kid. When he found out that mechanical engineers designed both, he knew he had found the job for him. Bill studied at Cornell University (he took astronomy from Carl Sagan!) and started working for airplane company Boeing. He even invented a part that helps quiet the noise of a 747 jet.

Bill's main contribution to science isn't as a scientist, though. His show *Bill Nye the Science Guy* introduced millions of kids (and adults) to the idea that science could be fun - even zany. In his trademark bowtie and long lab coat, Bill cut bread with a chainsaw, floated on a giant lily pad, and drove a grass-covered car to change the idea that scientists were boring, serious people doing boring, serious work.

Bill Nye the Science Guy won many awards and made Bill the new face of popular science in the US. Since then, Bill has hosted shows on the Discovery Channel, was the technical expert on *BattleBots*, and has shared the many ways science touches our lives on talk shows - and the importance of getting kids interested in it early. His famous light blue lab coat is on display in the National Museum of American History in his hometown of Washington, D.C.

Bill still loves to invent. He helped develop sundials for NASA's Mars rovers to check and adjust the cameras. (They were in some ways a tribute to his dad, who made his own sundials to keep track of time during World War II.) Bill holds four patents, from a magnifier made of water to an improved ballet shoe.

These days, Bill spends time trying to convince adults to pay attention to science, too. His Netflix show *Bill Nye Saves the World* was an adult look at vaccinations, the future of food, and other issues. He talks about climate change on TV and to groups around the country. He believes that to make the world better, we need to see and appreciate the science all around us.

BILL NYE

Born 1955

WASHINGTON, D.C. (USA)

MAE JEMISON

Born 1956

DECATUR, ALABAMA (USA)

MAE JEMISON

First woman of color to go to space, working to change the image of what a scientist looks like

Mae Jemison has been reaching for the stars since she was a little girl. Growing up in Chicago, where her family moved when she was 3, Mae loved watching the Apollo space missions on TV but was annoyed that there were no female astronauts. Partly inspired by *Star Trek*'s multicultural crew, unusual at the time, Mae decided that she would one day go to space.

Mae won a scholarship to Stanford University at 16, majoring in both chemical engineering and African and African American studies, followed by medical school at Cornell. Fluent in four languages (English, Japanese, Russian and Swahili), Mae joined the Peace Corps as a medical officer for two years in West Africa. When she returned to the United States, she worked as a doctor and applied to NASA. In 1987, Mae was one of only 15 selected out of over 2000 applications to the astronaut training program.

Mae finally got to go to space in 1992 aboard the space shuttle *Endeavour*, becoming the first woman of color in orbit. During the 8-day mission, Mae conducted experiments on bone cells and weightlessness. She started her shifts with a quote from her old favorite, *Star Trek* ("Hailing frequencies open").

Since leaving NASA, Mae has started several companies, taught at Dartmouth and Cornell, and started her own foundation. She leads the 100 Year Starship project, working to develop the technology to travel outside our solar system. She has written several books for kids and started The Earth We Share, an international science camp for teenagers and teachers.

Mae believes that diversity in science is not just nice but necessary, since more points of view help form better solutions to science's challenges. She is working to change the image of who does science, so that all kids – no matter where they're from or what they look like – know there is a place for them.

NEIL deGRASSE TYSON

Making astrophysics cool for a new generation

Neil deGrasse Tyson grew up in New York City, where light pollution makes it hard to see the stars. When he was 9, Neil visited the Hayden Planetarium – and was astonished at seeing the whole night sky. He felt like the universe was calling out to him. He has never lost that sense of wonder – and Neil not only grew up to be the Planetarium's director, he became one of the most famous scientists in America.

Neil was bitten hard by the astronomy bug. As a teenager, he hung out on the roof of his apartment building with his telescope. His college application to Cornell University showed so much enthusiasm that he got a personal invitation from Carl Sagan, who was a professor there and one of Neil's heroes (he ended up going to Harvard). Eventually Neil came back to New York to study astrophysics at Columbia University, completing his PhD in 1991.

Since then, Neil has become the popular face of science in the US, kind of a Carl Sagan for a new generation. He has been a frequent talk-show guest and hosted multiple TV series – including a revival of *Cosmos*, the series that made Carl a household name. Neil has published more than 16 books, not even counting his scientific research papers. He was a member of NASA's Advisory Council and two different groups advising the US president on science and space exploration. (He even has an asteroid named after him!)

Neil also uses social media to bring science to as many people as possible. He entertains millions of followers with fun science facts and by pointing out scientific errors in Hollywood movies. *StarTalk*, his radio show and podcast, uses comedy and pop culture to talk about astronomy, computers, and even the physics of sports. He wants us to understand just how much science affects every aspect of our lives – and how much fun making our own discoveries can be.

NEIL DEGRASSE TYSON

Born 1958

NEW YORK, NEW YORK (USA)

EMMANUELLE CHARPENTIER & JENNIFER DOUDNA

Revolutionized biology with a cheaper, easier, better way to edit genes

Since the double helix was discovered in the 1950s, scientists have been trying to make changes to DNA, the instruction manual for life. By deleting or adding genes, they hoped to cure diseases, create new plants, maybe even save endangered animals. But gene editing was slow, expensive, and not guaranteed to work.

That all changed in 2012, when scientists Jennifer Doudna and Emmanuelle Charpentier published a paper introducing CRISPR-Cas9 (pronounced "crisper"). CRISPR is like a cut-and-paste command, a molecular pair of scissors that scientists can use to remove and replace a specific gene. When the cells divide, they follow the new DNA instructions.

CRISPR itself isn't new. It was originally discovered to be a kind of immune system within bacteria, and scientists had been trying for several years to understand how it worked. In 2010, Emmanuelle showed how it could find and cut specific genomes. She met Jennifer soon after, and they decided to work together to see if CRISPR could be used as a scientific tool. From their labs in California and Sweden, they figured out how to use CRISPR to cut and swap DNA only where they wanted.

Their new method swept through labs around the world. CRISPR is easier, more accurate, and much cheaper than any of the old methods. Scientists saw so many possibilities: could it cure genetic diseases, even cancer? Create crops that need less water? All over the world, scientists are using CRISPR to look for new treatments and ways to adapt to our changing climate.

In 2020, Jennifer and Emmanuelle received the Nobel Prize in Chemistry – the first time a science Nobel was awarded to an all-female team. CRISPR has been one of the biggest stories in science for a decade, and we'll be studying its uses for years, looking for more ways to help people and the environment.

JENNIFER DOUDNA

Born 1964

WASHINGTON, D.C. (USA)

EMMANUELLE CHARPENTIER

Born 1968

JUVISY-SUR-ORGE (FRANCE)

PHILLIP ALVIOLA

Born 1977

LOS BAÑOS, LAGUNA (PHILIPPINES)

PHILLIP ALVIOLA

This real-life "Batman" is on the cutting edge of virus research and helps protect biodiversity

Phillip Alviola spent much of his childhood on Mount Makiling with his father, a wildlife biologist. This mystical mountain is named for a goddess said to protect the forest from those who would harm it – work that Phillip has also devoted his life to.

As a kid, Phillip loved the Indiana Jones movies about a globe-trotting archaeologist. While Indiana Jones usually hunted historical treasures, Phillip searches for knowledge about wildlife. His special interest is bats, one of the most diverse groups of mammals. He studies the ecology of their caves, the viruses they carry, and how conservation is connected to both.

When the COVID-19 crisis was traced to bats, Phillip's work became more urgent. His team of "virus hunters" studied bats all over the Philippines. They caught bats in wide nets, gently measured and swabbed them, then released them. Their data is being used to create models that might help predict – even avoid – future pandemics.

The Philippines is one of the most biodiverse countries in the world, and Phillip works to protect that diversity. He has helped discover new species of bats and other small forest mammals. For more than 10 years, Phillip and his colleagues traveled around the country, training national park staff and local people how to monitor biodiversity. His team's research helped get a national law passed to protect caves and their resources.

Phillip's work is known across Asia and around the world. He was honored in 2017 with a prestigious award from the Philippines' National Academy of Science and Technology. In 2021, he was one of only 28 scientists worldwide appointed to a new group at the World Health Organization to study and advise on new infectious diseases.

Phillip says that good research usually leads to more questions. "It's the eagerness, or sometimes the frustrations as well, to answer these questions that drives me to conduct more research."

MONKGOMOTSI MASENG

Born 1978

ORAPA, CENTRAL DISTRICT (BOTSWANA)

MONKGOMOTSI MASENG

Microbiologist and advocate for gender equity in science

When you take medicine, you expect it to work like it should. But every now and then, a person's genes cause medicine to work differently – or even not work at all. With a serious illness, that can be a matter of life and death.

In Botswana, Monkgomotsi Maseng is working to understand why this happens and how to fix it. Botswana has a lot of people living with HIV, a virus that attacks the immune system. Monkgomotsi studies how medicines used to treat HIV are processed in the body, looking for clues about why some people don't respond well or have more serious side effects. By understanding how genetics plays a role, she hopes to unlock ways that medicines could be personalized to work for an individual patient. Treatments for diabetes, high blood pressure, and other chronic illnesses could all be improved.

Monkgomotsi is a microbiologist, a scientist that studies living things too small to see with the naked eye, including bacteria and viruses. She even has a virus named after her – one she discovered in 2016 that can infect and kill the bacteria that causes tuberculosis. When the COVID-19 virus came to Botswana, Monkgomotsi used her experience to help figure out the best procedures. She also trained other scientists on how to use the equipment needed to detect the virus.

She is passionate about getting more women and girls into STEM fields. Growing up, she heard the stereotypes over and over again that science and math were just for boys. She started a program called Sister to Sister to make STEM fields more welcoming for everyone. Sister to Sister fights for more investment in education and provides mentors for young professionals and for students interested in going into STEM fields. She believes that making math and science more accessible to everyone can only help strengthen Botswana's social and economic growth.

CLARA BARKER

Tireless advocate for LGBTQ+ role models in science

Clara Barker loved science, but for a long time she wasn't the best student. Clara was assigned male at birth during a time when anti-gay laws in the UK made it hard to even talk about homosexuality or non-traditional families. Clara knew she was transgender very young, and life felt like a battle to keep going while hiding who she really was.

She eventually earned her PhD and became a material scientist. Material science uses physics, chemistry and engineering to develop new materials, like anti-scratch coatings for eyeglasses or touchscreen film that senses body heat. She loved coming up with new ways to do things.

But Clara almost left science. For many years, she met few openly gay scientists, and none who were openly trans. She saw hardly any women at the top levels of teaching and research. She was afraid that being herself might get her kicked out of her field. Once she started transitioning, though, she found many people accepting – they cared more about doing good science. She realized that her fears came from not having role models.

Now Clara tries to be that role model for others. She manages a lab at Oxford University and is involved with groups that advise the university on LGBTQ+ issues and work for more diversity and equality. In 2018, she was awarded Oxford's first diversity role model award. She works with groups for LGBTQ+ youth and often gives talks in schools about inclusion. In 2017, Clara won an award from the UK's prime minister in honor of her volunteer work.

Studies have shown that when LGBTQ+ people do not feel able to be themselves, the stress of hiding who they are keeps them from doing their best work. Science suffers as a result. Clara works hard to make STEM fields more welcoming for all – not only so her fellow scientists can live better lives, but so that they can do better science.

CLARA BARKER

Born 1979

READING, ENGLAND (UK)

OZAK ESU

Rising international star in engineering and advocate for women in the field

When Ozak Esu was growing up in Nigeria, the power would often go out when she was trying to watch cartoons. Frustrated, she asked her father whose job it was to fix the unreliable electricity. When he told her that electrical engineers would be responsible for that, she knew that was the path she wanted to follow.

At 17, Ozak moved to the UK to pursue her education and earned a PhD in electronic and electrical engineering. As a graduate student, she helped design electrical systems for more than 24 schools in just two years. She studied ways to make the electronics in wind turbines cheaper, so that wind power can be used in more places.

Ozak is passionate about sustainable building and renewable energy. Whether she is working on smart products for homes or hosting informational videos about the science of climate change, she is always looking for ways that engineering can make a difference in the world around us. She worked with a community in Malawi to help design and build a STEM education center there. She mentors young women studying in STEM fields in Nigeria and volunteers with groups trying to attract more women and people of color to engineering.

In 2017, Ozak won Young Woman Engineer of the Year from the UK's Institution of Engineering and Technology, and her name appears on many lists of exciting young engineers to watch. In a time when sustainable building is more urgent than ever, she is a rising star who believes that many problems in society can be addressed by engineering, if we just take a step back and look around us.

OZAK ESU

Born 1991

KADUNA (NIGERIA)

GRETA THUNBERG

Bringing worldwide attention to the science of climate change

Greta Thunberg, like most people, isn't a scientist. She was only 15 when she started skipping school on Fridays to sit outside the Swedish Parliament with a handwritten sign that read "School Strike for Climate." Yet there's no doubt that Greta has made waves in the scientific world. *Nature*, one of the world's top journals for science research, named her as one of their "Ten people who mattered to science" in 2019.

Scientific breakthroughs only change the world if people listen - and Greta gets people to listen. Her simple, direct protest has influenced millions of young people to organize in countries around the world. She has addressed the United Nations and argued with world leaders. The UK and Germany passed new laws to urgently reduce carbon emissions after Greta spoke there and helped local protesters put more pressure on their governments.

She does not pretend to have the answers on climate change - but she knows who does. "I don't want you to listen to me, I want you to listen to the scientists . . . and I want you to take real action," she told the US Congress. She is a link between science and politics, someone who stands up and demands that we listen to what scientists have to say. She is open about being on the autism spectrum, and credits that with her ability to stay focused on climate change and speak her mind honestly.

The world is taking notice. Greta has been nominated for the Nobel Peace Prize three times, and was the youngest person ever to be named *Time* magazine's Person of the Year. Climate scientists applaud her success in getting people to pay attention. This new awareness has motivated researchers to keep going in spite of slow government action. And she is inspiring many of tomorrow's environmental scientists. Though Greta may not be a scientist herself, she is definitely affecting the science of the future.

GRETA THUNBERG

Born 2003

STOCKHOLM (SWEDEN)

SCIENCE
PEOPLE

DAVID LEE CSICSKO

DAVID LEE CSICSKO is an award-winning artist and designer whose distinctive artwork, stained glass, and mosaics beautify train stations, hospitals, and universities across the Midwest and East Coast. His many credits include designing the Obamas' White House Christmas in 2012. David's lively illustrations can also be seen in *The Skin You Live In* from the Chicago Children's Museum, now in its 16th printing, and *LGBTQ+ Icons*, the first book of the People Series. Through his use of color, bold graphics and playful patterns, David Lee Csicsko celebrates the richness and diversity of life.

LINDY SINCLAIR is a producer and resident wordsmith for Trope Publishing Co.

Special thanks to **PATRICK McCARTHY** for editing assistance.

AERODYNAMICS The way air moves around things.

ALGEBRA Type of math that uses letters and symbols for missing pieces of information.

ANALYTICAL GEOMETRY A mathematical way to define shapes and get information about them.

ANATOMY Science that studies the structure of bodies, both inside and outside.

ANTHROPOLOGY The study of people and their cultures, both in the past and today.

ARCHAEOLOGY The study of things made and used by people in the past.

ASTRONOMY The study of outer space, including planets and stars.

ASTROPHYSICS The study of how stars and planets work, combining astronomy and physics.

AVIATION Anything to do with making or flying aircraft.

BACTERIOLOGY The study of bacteria, a type of single-cell organism.

BIODIVERSITY The variety of living things in a place.

BOTANY The study of plants.

CALCULUS Type of math that finds patterns and studies change.

CARTOGRAPHER A person who creates maps.

CHEMICAL ENGINEERING The use of chemical processes to solve problems or create useful products.

CHEMISTRY Major branch of science that studies what things are made of and how materials interact.

CHROMATOGRAPH A device that separates chemical mixtures into their parts.

CLIMATOLOGY The study of climates and how they change.

COMPOUND A chemical substance formed by atoms of different elements.

CONSERVATION The protection of nature and wise use of natural resources.

CYTOLOGY The study of cells, the small parts that make up all living things.

DISSECT To cut apart for examination.

ECOLOGY Science that studies the relationships between living things and their environment.

ECOSYSTEM A community of living things and their environment where each has a role to play.

ENGINEERING The use of physics, chemistry, and math to solve problems or invent new products.

ENTOMOLOGY The study of insects.

ENTREPRENEUR Someone who starts and runs a new business.

ENZYME A biological substance that speeds up a chemical reaction in the body.

EPILEPSY A condition that affects the electrical activity of the brain and causes seizures.

ETHNOBOTANY The study of how people of a particular place or culture use native plants.

EVOLUTION How living things change over a long period of time.

EXOBIOLOGY Science that studies whether there might be life in outer space.

EXPEDITION A journey with a specific purpose, usually for scientific research.

EXPERIMENTATION The process of trying out a new idea or activity.

EXTINCTION When a species dies out or disappears from earth.

GENES Small sections of DNA that contain specific bits of information.

GENOME An organism's complete set of DNA instructions.

GEOGRAPHY Science that studies the earth's surface and physical features.

GEOLOGY The study of rocks and minerals that make up the planet.

GEOMETRY A type of math that studies the shape, size, and position of things.

HERBIVORE An animal that only eats plants.

HERPETOLOGY The branch of zoology that studies reptiles and amphibians.

HYBRID A type of plant made by crossing two different plant varieties.

HYDROPOWER Electricity produced by moving water.

ICHTHYOLOGY The branch of zoology that studies fish.

ICHTHYOSAUR A giant marine reptile that looked kind of like a dolphin and lived millions of years ago.

MAGNETIC EQUATOR An imaginary line circling the earth, where a compass needle will stay balanced and not dip.

MAGNETIC STORMS A temporary disturbance in Earth's magnetic field, caused by unusual activity of the sun.

MALARIA A disease spread by mosquitos, most common in Africa and parts of Asia.

MARINE BIOLOGY The study of life in bodies of salt water, or organisms who depend on them.

METEOROLOGY The study of weather and the atmosphere.

MOLECULE The smallest possible particle of a substance, usually made of two or more atoms joined together.

NEUTRINO One of the most abundant particles in the universe, so tiny they are very hard to detect.

NOBEL PRIZE The world's highest awards in physics, chemistry, medicine, literature, and peace.

OMNIVORE An animal that eats both plants and meat.

PALEONTOLOGY The study of plants and animals that lived millions of years ago.

PATENT A document that gives an inventor the right to be the only one to make or sell a new invention for a certain number of years.

PENDULUM A weight that hangs from a string and can freely swing back and forth.

PHARMACOLOGY The study of how medicines are made and how they work.

PHILOSOPHY The study of the basic ideas about life, knowledge, and truth.

PHYSICS Major branch of science that studies energy, matter and motion.

PHYSIOLOGY A branch of biology that studies how living things work.

PLESIOSAUR A giant marine reptile that looked kind of like a turtle with a very long neck that lived millions of years ago.

POLYMATH A person who knows a lot about many different fields.

PULITZER PRIZE Annual prize in the US to honor great achievements in journalism, literature, and music.

QUANTUM PHYSICS The study of the smallest particles in the universe and how they behave.

QUARANTINE Staying away from other people to prevent spreading a disease.

QUASAR An area at the center of a distant galaxy that produces a huge amount of energy.

RADIOACTIVITY A type of energy given off by the decay of unstable atoms.

$E=mc^2$

RELATIVITY A scientific explanation of how space and time are related.

SCIENTIFIC METHOD A set of steps used to solve scientific problems and test new theories.

SONAR A way to detect things underwater using sound waves.

SUBATOMIC Anything smaller than an atom, including the particles an atom contains.

SUSTAINABILITY Using resources in a way so that they will continue to be available in the future.

TELEGRAPH A system of sending messages over long distances that existed before the telephone was invented.

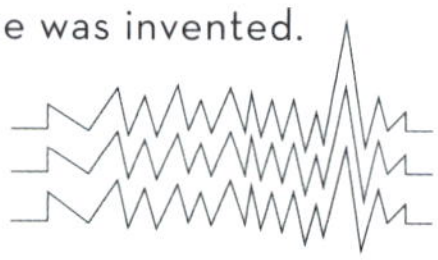

TUBERCULOSIS An infectious disease, mainly affecting the lungs, that was a leading cause of death in the 1800s.

VOLCANOLOGIST Someone who studies volcanoes.

X-RAY DIFFRACTION A way to study the structure of atoms by firing x-rays at a crystal, then observing the patterns created.

ZOOLOGY The study of animals and their behavior.

LCCN: 2022939758
ISBN: 978-1-9519630-7-1

Printed and bound in Turkey
First printing, 2022

David Lee Csicsko's illustrations from *Science People* are available for purchase. For inquiries, go to trope.com or email the gallery at info@trope.com.

+ INFORMATION:
For additional information on the Trope Edition Series, visit trope.com